SpringerBriefs in Computer Science

Series Editors
Stan Zdonik
Peng Ning
Shashi Shekhar
Jonathan Katz
XindongWu
Lakhmi C. Jain
David Padua
Xuemin Shen
Borko Furht
V.S. Subrahmanian
Martial Hebert
Katsushi Ikeuchi
Bruno Siciliano

T0222301

For further volumes:
http://www.springer.com/series/10028

Howard L. Weinert

Fast Compact Algorithms and Software for Spline Smoothing

 Springer

Howard L. Weinert
Johns Hopkins University
Baltimore, MD, USA

ISSN 2191-5768   ISSN 2191-5776 (electronic)
ISBN 978-1-4614-5495-3   ISBN 978-1-4614-5496-0 (eBook)
DOI 10.1007/978-1-4614-5496-0
Springer New York Heidelberg Dordrecht London

Library of Congress Control Number: 2012948342

Printed on acid-free paper

Springer is part of Springer Science+Business Media (www.springer.com)

For the Incomparable Cherie

Contents

1 Introduction 1
   1.1 The Continuous Problem 1
   1.2 The Solution 2
   1.3 Choosing the Smoothing Parameter 3
   1.4 A Look Ahead 4
   References 4

2 Cholesky Algorithm 5
   2.1 Normalized Cholesky Factorization 5
   2.2 Generalized Cross Validation Score 8
   2.3 MATLAB Implementation 11
   2.4 Monte Carlo Simulations 16
   References 18

3 QR Algorithm 19
   3.1 Condition Number of the Coefficient Matrix 19
   3.2 Least-Squares Formulation and QR Factorization 20
   3.3 MATLAB Implementation 23
   3.4 Monte Carlo Simulations 28
   References 28

4 FFT Algorithm 29
   4.1 Frequency Response of the Spline Smoother 29
   4.2 Computing the Spline 32
   4.3 Computing the GCV Score 32
   4.4 MATLAB Implementation 34
   4.5 Monte Carlo Simulations 35
   References 35

5   Discrete Spline Smoothing   37
   5.1   The Discrete Problem   37
   5.2   Cholesky Algorithm   38
   5.3   FFT Algorithm   42
   5.4   Discrete Versus Continuous   45
   References   45

Chapter 1
Introduction

Once upon a time, the most difficult aspect of signal processing was acquiring enough data. Nowadays, one can sample at very high rates and collect huge data sets, but processing the data in a timely manner, without exceeding available memory, is challenging, despite continuing advances in computer technology. Meeting this challenge requires the full exploitation of mathematical structure to develop an algorithm that can be implemented efficiently in software.

In this book, I investigate algorithmic alternatives for the problem of univariate cubic spline smoothing, a popular nonparametric curve fitting technique for extracting a signal from noise when there is little *a priori* information about either the signal or the noise. There are continuous and discrete versions of cubic spline smoothing. Discrete spline smoothing, also known as Whittaker-Henderson graduation, originated in 1899 [1] and was extensively studied in the first half of the twentieth century, primarily by actuarial mathematicians [2, 3, 4]. Most of its recent applications have been in economics, where it is called Hodrick-Prescott filtering [5]. Continuous spline smoothing was first investigated in the Sixties [6,7], and since then has largely pushed discrete spline smoothing into the background. Continuous splines have been applied in areas as diverse as tree-ring analysis [8], functional magnetic resonance imaging [9], and chromatography [10]. For further background, see [11–14].

1.1 The Continuous Problem

We have a set of uniformly spaced, noisy samples of a real signal $x(t)$, $t \in [0, 1]$,

$$y_i = x(i/n) + v_i, \quad i = 1, 2, \ldots, n, \tag{1.1}$$

and we want to estimate the signal at the sample locations. As estimates we will use samples $s(i/n)$ of the cubic smoothing spline $s(t)$ that minimizes

H.L. Weinert, *Fast Compact Algorithms and Software for Spline Smoothing*, SpringerBriefs in Computer Science, DOI 10.1007/978-1-4614-5496-0_1, © The Author(s) 2013

$$\xi \sum_{i=1}^{n} (y_i - u(i/n))^2 + \int_0^1 (u''(\tau))^2 d\tau, \tag{1.2}$$

among all $u(t)$ with a square-integrable second derivative. The spline is a cubic polynomial between each pair of adjacent sample locations. The positive parameter ξ determines the tradeoff between smoothness and fidelity to the data. When ξ is very small the spline samples will approximately lie on the least-squares regression line, and when ξ is very large the spline samples will be very close to the corresponding measurements. Note that since we are working on the interval [0,1], which we can do without loss of generality, the sampling rate is the number of measurements n.

It turns out [12] that the vector s of spline samples is the solution to a finite-dimensional minimization problem. Let M be the $(n - 2) \times n$ second difference matrix, and let P be the $(n - 2) \times (n - 2)$ symmetric, tridiagonal Toeplitz matrix with 2/3 and 1/6 on the diagonal and subdiagonal, respectively. When $n = 8$ for example,

$$M = \begin{bmatrix} 1 & -2 & 1 & 0 & 0 & 0 & 0 & 0 \\ 0 & 1 & -2 & 1 & 0 & 0 & 0 & 0 \\ 0 & 0 & 1 & -2 & 1 & 0 & 0 & 0 \\ 0 & 0 & 0 & 1 & -2 & 1 & 0 & 0 \\ 0 & 0 & 0 & 0 & 1 & -2 & 1 & 0 \\ 0 & 0 & 0 & 0 & 0 & 1 & -2 & 1 \end{bmatrix}, \tag{1.3}$$

$$P = \begin{bmatrix} 2/3 & 1/6 & 0 & 0 & 0 & 0 \\ 1/6 & 2/3 & 1/6 & 0 & 0 & 0 \\ 0 & 1/6 & 2/3 & 1/6 & 0 & 0 \\ 0 & 0 & 1/6 & 2/3 & 1/6 & 0 \\ 0 & 0 & 0 & 1/6 & 2/3 & 1/6 \\ 0 & 0 & 0 & 0 & 1/6 & 2/3 \end{bmatrix}. \tag{1.4}$$

Then s minimizes

$$\lambda(y - u)^T (y - u) + u^T M^T P^{-1} M u, \tag{1.5}$$

among all vectors u, where y is the vector of measurements and $\lambda = \xi/n^3$. Note that M is full rank and P is positive definite.

1.2   The Solution

The minimizer s of (1.5) satisfies

$$(\lambda I + M^T P^{-1} M)s = \lambda y. \tag{1.6}$$

Although this coefficient matrix is positive definite, persymmetric, and even quasi-Toeplitz, there is an alternative equation with a coefficient matrix that has more exploitable structure. Multiplying (1.6) on the left by M and letting $c = \lambda^{-1}P^{-1}Ms$, we obtain

$$\left(\lambda P + MM^T\right)c = My,\tag{1.7}$$

$$s = y - M^T c.\tag{1.8}$$

The coefficient matrix $A = \lambda P + MM^T$ is now positive definite, Toeplitz, and banded (pentadiagonal). When $n = 8$ for example,

$$A = \begin{bmatrix} \frac{2}{3}\lambda+6 & \frac{1}{6}\lambda-4 & 1 & 0 & 0 & 0 \\ \frac{1}{6}\lambda-4 & \frac{2}{3}\lambda+6 & \frac{1}{6}\lambda-4 & 1 & 0 & 0 \\ 1 & \frac{1}{6}\lambda-4 & \frac{2}{3}\lambda+6 & \frac{1}{6}\lambda-4 & 1 & 0 \\ 0 & 1 & \frac{1}{6}\lambda-4 & \frac{2}{3}\lambda+6 & \frac{1}{6}\lambda-4 & 1 \\ 0 & 0 & 1 & \frac{1}{6}\lambda-4 & \frac{2}{3}\lambda+6 & \frac{1}{6}\lambda-4 \\ 0 & 0 & 0 & 1 & \frac{1}{6}\lambda-4 & \frac{2}{3}\lambda+6 \end{bmatrix}.\tag{1.9}$$

The vectors s and c can be used to compute spline values (signal estimates) between adjacent measurement locations [12,14], but for high sampling rates these interpolated values are generally not needed.

1.3   Choosing the Smoothing Parameter

Eq. (1.7) cannot be solved until a value is assigned to the smoothing parameter λ. Ideally, λ should be chosen to minimize the true mean square error, but this is impossible since the signal is unknown. Instead, we will choose λ to minimize the generalized cross validation (GCV) score

$$GCV(\lambda) = \frac{n^{-1}c^T MM^T c}{\left(n^{-1}\text{trace}\left(M^T A^{-1} M\right)\right)^2}.\tag{1.10}$$

For large n, minimizing (1.10) approximately minimizes the true mean square error. Generalized cross validation [11,12] is the most popular of several automatic adaptive procedures for choosing the smoothing parameter [15]. The best option for minimizing (1.10) is Brent's method [16], which is a combination of golden section search and parabolic interpolation. Since the GCV score must be computed for each trial value of λ, algorithm efficiency is very important.

1.4 A Look Ahead

The next three chapters present algorithms for computing cubic smoothing splines with generalized cross validation. The algorithms are based on Cholesky factorization, QR factorization, and the fast Fourier transform (FFT). Chapter 5 deals with discrete splines. Each chapter includes the relevant software, as well as simulation results and algorithm comparisons. All programs were written in MATLAB (R2012a, 64 bit) and executed on a Dell VOSTRO 3750 laptop with an Intel Core i5-2410 M CPU (2.3 GHz) running Windows 7 Professional (64 bit).

References

[1] Bohlmann G (1899) Ein ausgleichungsproblem. Nachr Ges Wiss Gott Math-Phys Kl:260–271
[2] Whittaker ET (1923) On a new method of graduation. Proc Edinb Math Soc 41:63–75
[3] Henderson R (1925) Further remarks on graduation. Trans Actuar Soc Am 26:52–57
[4] Spoerl CA (1937) The Whittaker-Henderson graduation formula A. Trans Actuar Soc Am 38:403-462
[5] Hodrick RJ, Prescott EC (1997) Postwar US business cycles: an empirical investigation. J Money Credit Bank 29:1–16
[6] Schoenberg IJ (1964) Spline functions and the problem of graduation. Proc Natl Acad Sci 52:947–950
[7] Reinsch CH (1967) Smoothing by spline functions. Numer Math 10:177–183
[8] Cook ER, Peters K (1981) The smoothing spline: a new approach to standardizing forest interior tree-ring width series for dendroclimatic studies. Tree-Ring Bull 41:45–53
[9] Carew JD, Wahba G, Xie X, Nordheim EV, Meyerand ME (2003) Optimal spline smoothing of fMRI time series by generalized cross-validation. NeuroImage 18:950–961
[10] Kuligowski J, Carrion D, Quintas G, Garrigues S, de la Guardia M (2010) Cubic smoothing splines background correction in on-line liquid chromatography-Fourier transform infrared spectrometry. J Chromatogr. A 1217:6733–6741
[11] Wahba G (1990) Spline models for observational data. SIAM, Philadelphia
[12] Green PJ, Silverman BW (1994) Nonparametric regression and generalized linear models. Chapman and Hall, London
[13] Weinert HL (2007) Efficient computation for Whittaker-Henderson smoothing. Comput Stat Data Anal 52:959–974
[14] Weinert HL (2009) A fast compact algorithm for cubic spline smoothing. Comput Stat Data Anal 53:932–940
[15] Lee TCM (2003) Smoothing parameter selection for smoothing splines: a simulation study. Comput Stat Data Anal 42:139–148
[16] Brent RP (1973) Algorithms for minimization without derivatives. Prentice-Hall, Englewood Cliffs

Chapter 2
Cholesky Algorithm

Reinsch [1] provided the first practical algorithm for the continuous case. He solved (1.7)-(1.8) with O(n) floating point operations (flops) using a normalized Cholesky factorization of the coefficient matrix, with a predetermined value for the smoothing parameter. Hutchinson and de Hoog [2] showed that the GCV score could also be evaluated with O(n) flops. However, both execution time and memory use can be reduced substantially by digging deeper into the structure of the problem.

2.1 Normalized Cholesky Factorization

The coefficient matrix in (1.7) can be factored as

$$A = LDL^T, \tag{2.1}$$

where L is unit lower triangular and banded, and D is diagonal with positive diagonal entries. When $n = 8$ for example,

$$
L = \begin{bmatrix}
1 & 0 & 0 & 0 & 0 & 0 \\
-e_1 & 1 & 0 & 0 & 0 & 0 \\
f_1 & -e_2 & 1 & 0 & 0 & 0 \\
0 & f_2 & -e_3 & 1 & 0 & 0 \\
0 & 0 & f_3 & -e_4 & 1 & 0 \\
0 & 0 & 0 & f_4 & -e_5 & 1
\end{bmatrix}, \tag{2.2}
$$

H.L. Weinert, *Fast Compact Algorithms and Software for Spline Smoothing*,
SpringerBriefs in Computer Science, DOI 10.1007/978-1-4614-5496-0_2,
© The Author(s) 2013

$$D = \begin{bmatrix} f_1^{-1} & 0 & 0 & 0 & 0 & 0 \\ 0 & f_2^{-1} & 0 & 0 & 0 & 0 \\ 0 & 0 & f_3^{-1} & 0 & 0 & 0 \\ 0 & 0 & 0 & f_4^{-1} & 0 & 0 \\ 0 & 0 & 0 & 0 & f_5^{-1} & 0 \\ 0 & 0 & 0 & 0 & 0 & f_6^{-1} \end{bmatrix}. \tag{2.3}$$

As we compute the factorization we can solve the triangular system

$$LD\theta = My. \tag{2.4}$$

There are many ways to arrange the computations but the following requires the fewest flops. First evaluate the right side of (2.4) with

$$\delta_i = y_{i+1} - y_i, \quad i = 1, 2, \ldots, n-1, \tag{2.5}$$

and

$$w_i = \delta_{i+1} - \delta_i, \quad i = 1, 2, \ldots, n-2. \tag{2.6}$$

Then

$$a_0 = 6 + \frac{2}{3}\lambda, \quad a_1 = 4 - \frac{1}{6}\lambda,$$

$$f_1 = \frac{1}{a_0}, \quad \theta_1 = f_1 w_1, \quad \mu_1 = a_1, \quad e_1 = \mu_1 f_1,$$

$$f_2 = \frac{1}{a_0 - \mu_1 e_1}, \quad \theta_2 = f_2(w_2 + \mu_1 \theta_1), \quad \mu_2 = a_1 - e_1, \quad e_2 = \mu_2 f_2, \tag{2.7}$$

and for $i = 3, 4, \ldots, n-2$,

$$f_i = \frac{1}{a_0 - \mu_{i-1} e_{i-1} - f_{i-2}},$$
$$\theta_i = f_i(w_i + \mu_{i-1}\theta_{i-1} - \theta_{i-2}),$$
$$\mu_i = a_1 - e_{i-1}, \quad e_i = \mu_i f_i. \tag{2.8}$$

The next step is to solve the triangular system

$$L^T c = \theta \tag{2.9}$$

as follows:

$$c_{n-2} = \theta_{n-2}, \quad c_{n-3} = \theta_{n-3} + e_{n-3}c_{n-2}, \tag{2.10}$$

$$c_i = \theta_i + e_i c_{i+1} - f_i c_{i+2}, \quad i = n-4, n-5, \ldots, 1. \tag{2.11}$$

Then evaluate $M^T c$ in (1.8) with

$$\phi_1 = 0, \quad \phi_2 = c_1, \quad \phi_i = c_{i-1} - c_{i-2}, \quad \phi_n = -c_{n-2}, \quad \phi_{n+1} = 0, \qquad (2.12)$$

for $i = 3, 4, \ldots, n - 1$, and

$$\psi_i = \phi_{i+1} - \phi_i, \quad i = 1, 2, \ldots, n. \qquad (2.13)$$

Finally,

$$s_i = y_i - \psi_i, \quad i = 1, 2, \ldots, n. \qquad (2.14)$$

For large n, this method of solving (1.7)-(1.8) requires about $18n$ additions and multiplications and n divisions. Since a division typically takes fifty percent longer than an addition or multiplication, it is important to keep the number of divisions as small as possible. In the MATLAB implementation we will conserve memory by overwriting δ with w, θ with c, each μ_i with μ_{i+1}, c with ϕ, ϕ with ψ, and ψ with s. As a result, only about $40n$ bytes of memory will be needed.

We can further reduce execution time and memory use by using a little known result [3] about convergence along the diagonals of L and D. Consider the palindromic polynomial whose coefficients are the entries in the rows of A:

$$z^4 + (\lambda/6 - 4)z^3 + (2\lambda/3 + 6)z^2 + (\lambda/6 - 4)z + 1. \qquad (2.15)$$

Its roots occur in reciprocal pairs with two inside the unit circle and two outside. If z_1 and z_2 denote the roots inside the unit circle, then as $i, n \to \infty$, $e_i \to e_0$ and $f_i \to f_0$ at the same rate, where

$$e_0 = z_1 + z_2, \quad f_0 = z_1 z_2. \qquad (2.16)$$

Consequently, if we compute the e_i and f_i only until they converge to within machine precision (unit roundoff of 1.1×10^{-16}), the number N of required iterations is

$$N = \text{ceil} \left(\frac{\log_{10}(1.1) - 16}{2\log_{10}(\rho)} \right), \qquad (2.17)$$

where $\rho = \max(|z_1|, |z_2|)$.

Eq. (2.17) allows us to determine in advance the number of iterations required to carry out the Cholesky factorization, so that we don't have to test for convergence at each iteration. However, we must still find polynomial roots for each choice of λ. We can avoid that task by replacing (2.17) with a simple formula that expresses the number of required iterations directly in terms of λ. First, one can determine empirically that $\log_{10}(N)$ depends almost linearly on $\log_{10}(\lambda)$ for $\lambda \leq 100$, and

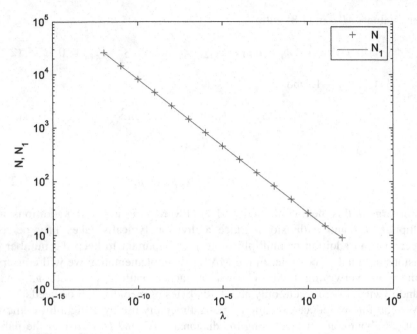

Fig. 2.1 Number of iterations for convergence

Table 2.1 Number of iterations for convergence

λ	10^{-12}	10^{-10}	10^{8}	10^{-6}	10^{-4}	10^{-2}	1	100
N	25973	8215	2598	822	260	83	26	9
N_1	26000	8222	2600	823	260	83	26	9

that $9 \leq N \leq 14$ for $\lambda > 100$. A formula for the approximate number N_1 of required iterations is thus

$$N_1 = \begin{cases} \text{ceil}\left(26\lambda^{-1/4}\right), & \lambda \leq 100 \\ 14, & \lambda > 100 \end{cases}. \tag{2.18}$$

Fig. 2.1 and Table 2.1 compare N and N_1 for $\lambda \leq 100$. If we exploit the convergence, then for large n, solving (1.7)-(1.8) requires about $13n + 5N_1$ additions and multiplications and N_1 divisions, and $24n + 16N_1$ bytes of memory.

2.2 Generalized Cross Validation Score

Because M is sparse, the trace in (1.10) depends on only a few entries in A^{-1}. If g_i, d_i, and p_i respectively denote the entries on the diagonal and first and second superdiagonals of A^{-1}, then

$$\text{trace}\left(M^T A^{-1} M\right) = 6 \sum_{i=1}^{n-2} g_i - 8 \sum_{i=1}^{n-3} d_i + 2 \sum_{i=1}^{n-4} p_i. \tag{2.19}$$

The time required to compute this trace can be reduced by about fifty percent because the persymmetry of A^{-1} means that the g_i, d_i, and p_i sequences are symmetric. When $n = 8$ for example,

$$A^{-1} = \begin{bmatrix} g_1 & d_1 & p_1 & x & x & x \\ d_1 & g_2 & d_2 & p_2 & x & x \\ p_1 & d_2 & g_3 & d_3 & p_2 & x \\ x & p_2 & d_3 & g_3 & d_2 & p_1 \\ x & x & p_2 & d_2 & g_2 & d_1 \\ x & x & x & p_1 & d_1 & g_1 \end{bmatrix}. \tag{2.20}$$

Therefore, (2.19) can be replaced by

$$\text{trace}\left(M^T A^{-1} M\right) = 6\left(2 \sum_{l=1}^{n'-1} g_i - \alpha g_{n'-1}\right)$$

$$- 8\left(2 \sum_{i=1}^{n'-1} d_i - (\alpha+1)d_{n'-1}\right)$$

$$+ 2\left(2 \sum_{i=1}^{n'-2} p_i - \alpha p_{n'-2}\right), \tag{2.21}$$

where

$$n' = \begin{cases} \dfrac{n}{2}, & n \text{ even} \\ \dfrac{n+1}{2}, & n \text{ odd} \end{cases}, \tag{2.22}$$

$$\alpha = \begin{cases} 0, & n \text{ even} \\ 1, & n \text{ odd} \end{cases}. \tag{2.23}$$

To compute just the required entries in A^{-1}, start with the identity

$$L^T A^{-1} = D^{-1} L^{-1}. \tag{2.24}$$

Note that the matrix on the right hand side is lower triangular with the f_i on the diagonal. Equate the $(n-2, n-2)$ entries on both sides to get an equation for g_1. Then equate the $(n-3, n-2)$ entries to get d_1, the $(n-3, n-3)$ entries to get g_2, the $(n-4, n-2)$

entries to get p_1, the $(n\text{-}4, n\text{-}3)$ entries to get d_2, and so on. The resulting equations are

$$g_1 = f_{n-2}, \quad d_1 = e_{n-3}g_1, \quad g_2 = f_{n-3} + e_{n-3}d_1, \tag{2.25}$$

and for $i = 3, 4, \ldots, n' - 1$,

$$
\begin{aligned}
p_{i-2} &= e_{n-i-1}d_{i-2} - f_{n-i-1}g_{i-2}, \\
d_{i-1} &= e_{n-i-1}g_{i-1} - f_{n-i-1}d_{i-2}, \\
g_i &= e_{n-i-1}d_{i-1} + (1 - p_{i-2})f_{n-i-1},
\end{aligned}
\tag{2.26}
$$

and finally,

$$
\begin{aligned}
p_{n'-2} &= e_{n-n'-1}d_{n'-2} - f_{n-n'-1}g_{n'-2}, \\
d_{n'-1} &= e_{n-n'-1}g_{n'-1} - f_{n-n'-1}d_{n'-2}.
\end{aligned}
\tag{2.27}
$$

Evaluating the GCV score (1.10) using (2.21) and (2.25)-(2.27) requires about $8.5n$ additions and multiplications for large n. Only a trivial amount of memory is needed since we can successively overwrite each g_i, d_i, p_i, and accumulate the sums in (2.21) as we proceed.

Interestingly, the g_i, d_i, and p_i sequences also converge to within machine precision in N (or N_1) iterations. The theoretical limits, obtained from (2.26), are

$$
\begin{aligned}
g_0 &= \frac{f_0(1 + f_0)}{(1 - f_0)\left((1 + f_0)^2 - e_0^2\right)}, \\
d_0 &= \frac{e_0 f_0}{(1 - f_0)\left((1 + f_0)^2 - e_0^2\right)}, \\
p_0 &= \frac{e_0^2 f_0 - (1 + f_0)f_0^2}{(1 - f_0)\left((1 + f_0)^2 - e_0^2\right)}.
\end{aligned}
\tag{2.28}
$$

As long as $N_1 \leq n' - 2$, we can truncate the iterations in (2.26) to reduce computation time. The GCV score evaluation will then require only about $2n + 13N_1$ additions and multiplications. In all, the computation of the spline vector and the GCV score for large n and for a single value of λ entails at most $26.5n$ additions and multiplications and n divisions, and at least $15n + 18N_1$ additions and multiplications and N_1 divisions.

When the GCV score is minimized, most of the computations must be repeated for each new choice of λ. If m values of λ are used, we will need at most $3n + 23.5mn$ additions and multiplications and mn divisions, and at least

$3n + 12mn + 18mN_1$ additions and multiplications and mN_1 divisions to compute the spline vector corresponding to the optimal λ. Typically with Brent's method, $10 \leq m \leq 20$. When $n \gg N_1$ truncating the e_i, f_i, g_i, d_i, p_i iterations reduces the total flop count by about 50 % without affecting the signal estimation accuracy. Only $24n$ bytes of memory are required in this case.

2.3 MATLAB Implementation

The Cholesky-based algorithm is implemented in a MATLAB function named *splinechol*. The input is the column vector of measurements y. The output is the column vector of spline values s. The minimization of the GCV score is carried out by the MATLAB function *fminbnd* which is based on Brent's method. The minimization is faster and more accurate if *fminbnd* searches for the best σ instead of the best λ, where

$$\lambda - \frac{4\sigma^4}{1 - \sigma^2}. \tag{2.29}$$

Note that $\sigma \in (0, 1)$. All computations that must be repeated when σ is changed are in the nested function *gcv*. After *fminbnd* terminates, an extra call to *gcv* is necessary to compute, using the optimal σ, those quantities that determine the spline.

```
function s = splinechol(y)
%
n = length(y);
nc = ceil(n/2);
rmndr = rem(n,2);
w = diff(y,2);
sig = fminbnd(@gcv, 0, 1);
gcv(sig);
s = y-s;
%
    function score = gcv(sig)
%
    s = zeros(n-2,1);
    lam = 4*sig^4/(1-sig^2);
    a0 = 6+lam*2/3;
    a1 = 4-lam/6;
    if lam > 100
      N = 14;
    else
      N = ceil(26*lam^-.25);
```

```
  end
  if N > n-5 % No truncation
    e = zeros(1,n-2);
    f = zeros(1,n-2);
    f(1) = 1/a0;
    s(1) = f(1)*w(1);
    e(1) = a1*f(1);
    f(2) = 1/(a0-a1*e(1));
    s(2) = f(2)*(w(2)+a1*s(1));
    mu = a1-e(1);
    e(2) = mu*f(2);
    for k = 3:n-2
      f(k) = 1/(a0-mu*e(k-1)-f(k-2));
      s(k) = f(k)*(w(k)+mu*s(k-1)-s(k-2));
      mu = a1-e(k-1);
      e(k) = mu*f(k);
    end
    s(n-3) = s(n-3)+e(n-3)*s(n-2);
    for k = n-4:-1:1
      s(k) = s(k)+e(k)*s(k+1)-f(k)*s(k+2);
    end
    g2 = f(n-2);
    tr1 = g2;
    d = e(n-3)*g2;
    tr2 = d;
    g1 = f(n-3)+e(n-3)*d;
    tr1 = tr1+g1;
    tr3 = 0;
    for k = n-4:-1:n-nc
      p = e(k)*d-f(k)*g2;
      tr3 = tr3+p;
      d = e(k)*g1-f(k)*d;
      tr2 = tr2+d;
      g2 = g1;
      g1 = f(k)*(1-p)+e(k)*d;
      tr1 = tr1+g1;
    end
    p = e(n-nc-1)*d-f(n-nc-1)*g2;
    tr3 = tr3+p;
    d = e(n-nc-1)*g1-f(n-nc-1)*d;
    tr2 = tr2+d;
  else % Truncate e,f iterations
    e = zeros(1,N);
    f = zeros(1,N);
    f(1) = 1/a0;
```

```
s(1) = f(1)*w(1);
e(1) = a1*f(1);
f(2) = 1/(a0-a1*e(1));
s(2) = f(2)*(w(2)+a1*s(1));
mu = a1-e(1);
e(2) = mu*f(2);
for k = 3:N
  f(k) = 1/(a0-mu*e(k-1)-f(k-2));
  s(k) = f(k)*(w(k)+mu*s(k-1)-s(k-2));
  mu = a1-e(k-1);
  e(k) = mu*f(k);
end
flim = f(N);
elim = e(N);
mu = a1-elim;
for k = N+1:n-2
  s(k) = flim*(w(k)+mu*s(k-1)-s(k-2));
end
s(n-3) = s(n-3)+elim*s(n-2);
for k = n-4:-1:N
  s(k) = s(k)+elim*s(k+1)-flim*s(k+2);
end
for k = N-1:-1:1
  s(k) = s(k)+e(k)*s(k+1)-f(k)*s(k+2);
end
g2 = flim;
tr1 = g2;
d = elim*g2;
tr2 = d;
g1 = flim+elim*d;
tr1 = tr1+g1;
tr3 = 0;
if N < nc-1 % Truncate g,d,p iterations
  for k = 3:N
    p = elim*d-flim*g2;
    tr3 = tr3+p;
    d = elim*g1-flim*d;
    tr2 = tr2+d;
    g2 = g1;
    g1 = flim*(1-p)+elim*d;
    tr1 = tr1+g1;
  end
  tr1 = tr1+(nc-N-1)*g1;
  tr2 = tr2+(nc-N)*d;
  tr3 = tr3+(nc-N)*p;
```

```
    else % Don't truncate g,d,p iterations
      for k = n-4q:-1:N
        p = elim*d-flim*g2;
        tr3 = tr3+p;
        d = elim*g1-flim*d;
        tr2 = tr2+d;
        g2 = g1;
        g1 = flim*(1-p)+elim*d;
        tr1 = tr1+g1;
      end
      for k = N-1:-1:n-nc
         p = e(k)*d-f(k)*g2;
         tr3 = tr3+p;
         d = e(k)*g1-f(k)*d;
         tr2 = tr2+d;
         g2 = g1;
         g1 = f(k)*(1-p)+e(k)*d;
         tr1 = tr1+g1;
      end
      p = e(n-nc-1)*d-f(n-nc-1)*g2;
      tr3 = tr3+p;
      d = e(n-nc-1)*g1-f(n-nc-1)*d;
      tr2 = tr2+d;
    end
  end
  tr = 6*(2*tr1-rmndr*g1)-8*(2*tr2-(1+rmndr)*d)...
                        +2*(2*tr3-rmndr*p);
  s = diff([0; 0; s; 0; 0],2);
  score = n*(s'*s)/tr^2;
  end
end
```

For comparison purposes, I also created a stripped-down version of *splinechol*, named *csplineopt*, which does not use iteration truncation.

```
function s = csplineopt(y)
%
n = length(y);
nc = ceil(n/2);
rmndr = rem(n,2);
w = diff(y,2);
sig = fminbnd(@cgcv, 0, 1);
cgcv(sig);
s = y-s;
  function score = cgcv(sig)
  s = zeros(n-2,1);
```

```
lam = 4*sig^4/(1-sig^2);
a0 = 6+lam*2/3;
a1 = 4-lam/6;
e = zeros(1,n-2);
f = zeros(1,n-2);
f(1) = 1/a0;
s(1) = f(1)*w(1);
e(1) = a1*f(1);
f(2) = 1/(a0-a1*e(1));
s(2) = f(2)*(w(2)+a1*s(1));
 mu = a1-e(1);
e(2) = mu*f(2);
for j = 3:n-2
  f(j) = 1/(a0-mu*e(j-1)-f(j-2));
  s(j) = f(j)*(w(j)+mu*s(j-1)-s(j-2));
  mu = a1-e(j-1);
  e(j) = mu*f(j);
end
s(n-3) = s(n-3)+e(n-3)*s(n-2);
for j = n-4:-1:1
  s(j) = s(j)+e(j)*s(j+1)-f(j)*s(j+2);
end
g2 = f(n-2);
d = e(n-3)*g2;
tr2 = d;
g1 = f(n-3)+e(n-3)*d;
tr1 = g2+g1;
tr3 = 0;
for j = n-4:-1:n-nc
  p = e(j)*d-f(j)*g2;
  tr3 = tr3+p;
  d = e(j)*g1-f(j)*d;
  tr2 = tr2+d;
  g2 = g1;
  g1 = f(j)*(1-p)+e(j)*d;
  tr1 = tr1+g1;
end
p = e(n-nc-1)*d-f(n-nc-1)*g2;
d = e(n-nc-1)*g1-f(n-nc-1)*d;
tr = 6*(2*tr1-rmndr*g1)-8*(2*tr2+(1-rmndr)*d)...
+2*(2*tr3+(2-rmndr)*p);
s = diff([0; 0; s; 0; 0],2);
score = n*(s'*s)/tr^2;
end
end
```

Finally, *splinesimple* neither computes nor minimizes the GCV score and does not truncate the iterations. The user must input a specific λ.

```
function s = splinesimple(y,lam)
%
n = length(y);
s = zeros(n-2,1);
w = diff(y,2);
a0 = 6+lam*2/3;
a1 = 4-lam/6;
e = zeros(1,n-2);
f = zeros(1,n-2);
f(1) = 1/a0;
s(1) = f(1)*w(1);
e(1) = a1*f(1);
f(2) = 1/(a0-a1*e(1));
s(2) = f(2)*(w(2)+a1*s(1));
mu = a1-e(1);
e(2) = mu*f(2);
for j = 3:n-2
   f(j) = 1/(a0-mu*e(j-1)-f(j-2));
   s(j) = f(j)*(w(j)+mu*s(j-1)-s(j-2));
   mu = a1-e(j-1);
   e(j) = mu*f(j);
end
s(n-3) = s(n-3)+e(n-3)*s(n-2);
for j = n-4:-1:1
   s(j) = s(j)+e(j)*s(j+1)-f(j)*s(j+2);
end
s = diff([0; 0; s; 0; 0],2);
s = y-s;
```

2.4 Monte Carlo Simulations

First, *splinechol* will be compared to *csplineopt* in terms of speed and accuracy. This will show the advantage to be gained by allowing truncation of the e_i, f_i, g_i, d_i, p_i iterations. Measurements were generated as in (1.1) by sampling three different signals:

$$
\begin{aligned}
x_1(t) &= 2 + \sin(2200\pi t),\\
x_2(t) &= 2 + 0.3e^{-64(t-0.25)^2} + 0.7e^{-256(t-0.75)^2},\\
x_3(t) &= 4 - 48t + 218t^2 - 315t^3 + 145t^4.
\end{aligned}
\tag{2.30}
$$

Table 2.2 RMS error for n = 10^6

signal	SNR = 20 dB splinechol	SNR = 20 dB csplineopt	SNR = 40 dB splinechol	SNR = 40 dB csplineopt
x_1	1.7×10^{-2}	1.7×10^{-2}	2.2×10^{-3}	2.2×10^{-3}
x_2	4.4×10^{-3}	6.1×10^{-3}	2.4×10^{-4}	6.3×10^{-4}
x_3	3.5×10^{-3}	8.9×10^{-3}	3.6×10^{-4}	9.0×10^{-4}

The noise vector added to the vector of signal samples was $v = \beta r$, where $\beta > 0$ and $r = randn(n, 1)$. The MATLAB function *randn* generates pseudorandom values from a standard normal distribution. The parameter β was chosen to produce a specific signal-to-noise ratio (SNR). Since

$$SNR = 10\log_{10}\left(\frac{x^T x}{\beta^2 r^T r}\right), \tag{2.31}$$

we have

$$\beta = 10^{-SNR/20}\sqrt{\frac{x^T x}{r^T r}}. \tag{2.32}$$

For each signal, two values of n (10^4, 10^6) and two values of SNR (20 dB, 40 dB) were used. Execution time was measured with MATLAB's *tic* and *toc* functions, and the RMS error (RMSE) was used to determine accuracy:

$$RMSE = \sqrt{\frac{1}{n}(s - x)^T(s - x)}. \tag{2.33}$$

With the first signal in (2.30), *splinechol* and *csplineopt* had the same RMSE; for the other two signals, *splinechol* and *csplineopt* had the same RMSE for $n = 10^4$, but *splinechol* was more accurate for $n = 10^6$. See Table 2.2. In all cases, *splinechol* was about twice as fast as *csplineopt* and required as little as 60% of the memory. We can conclude that allowing truncation of the e_i, f_i, g_i, d_i, p_i iterations significantly reduces execution time and memory use without compromising accuracy.

The best commercially available software for computing cubic smoothing splines is the function *csaps* in MATLAB's Curve Fitting Toolbox. This function is based on a normalized Cholesky factorization and takes account of the sparseness of the coefficient matrix to reduce memory use, but it does not attempt to find the best smoothing parameter. Instead, the user must input a value for a normalized parameter $p \in (0, 1)$, where p is related to λ via

$$p = \frac{n^3 \lambda}{n^3 \lambda + 1}.$$
(2.34)

The function *splinesimple*, which also produces the spline for a single value of the smoothing parameter, is much more efficient. In fact, *splinesimple* is 20 times faster than *csaps* and uses only 11 % of the memory, without degrading estimation accuracy. The function *splinechol*, which does much more than *csaps*, is still about twice as fast and uses only 7-11 % of the memory.

References

[1] Reinsch CH (1967) Smoothing by spline functions. Numer Math 10:177–183
[2] Hutchinson MF, de Hoog FR (1985) Smoothing noisy data with spline functions. Numer Math 47:99–106
[3] Bauer FL (1955) Ein direktes iterationsverfahren zur Hurwitz-zerlegung eines polynoms. Arch Elektr Ubertragung 9:285–290

Chapter 3
QR Algorithm

The coefficient matrix A (1.9) in the normal equations (1.7) will be ill-conditioned for small λ, causing the number of correct digits in the computed spline to be small. To try to compensate for this problem, one can reformulate spline smoothing as a basic least-squares problem and solve it using a QR factorization. De Hoog and Hutchinson [1], building on earlier work [2, 3, 4] on general banded least squares problems, presented a QR algorithm for spline smoothing. In this chapter we will evaluate the condition number of the coefficient matrix, present a faster and more compact QR algorithm, and determine whether this alternative is preferable to solving the normal equations.

3.1 Condition Number of the Coefficient Matrix

The (2-norm) condition number of A, denoted $\kappa(A)$, is the ratio of the largest singular value to the smallest. As $\kappa(A)$ increases, the perturbation sensitivity of (1.7) increases. For each order of magnitude increase in $\kappa(A)$, there is one less correct digit in the computed spline [5, 6, 7]. While there is no exact formula for $\kappa(A)$, an approximate formula can be found. First note that

$$\lim_{\lambda \to 0} \kappa(A) = \kappa(MM^T),$$

$$\lim_{\lambda \to \infty} \kappa(A) = \kappa(P). \tag{3.1}$$

Using the *cond* function in MATLAB, one can determine that

$$\kappa(MM^T) \cong 0.032n^4,$$

$$\kappa(P) \leq 3, \quad \text{for all } n, \tag{3.2}$$

H.L. Weinert, *Fast Compact Algorithms and Software for Spline Smoothing*, SpringerBriefs in Computer Science, DOI 10.1007/978-1-4614-5496-0_3, © The Author(s) 2013

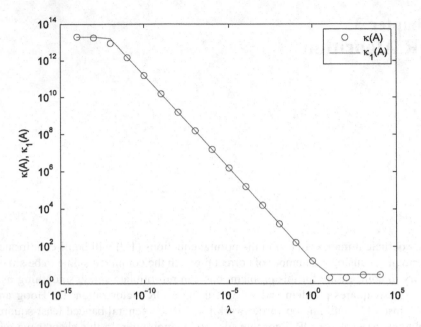

Fig. 3.1 Condition number for $n = 5000$

and for a wide range of intermediate λ values, $\log_{10}\kappa(A)$ is approximately a linear function of $\log_{10}(\lambda)$. An approximate condition number $\kappa_1(A)$ is given by

$$\kappa_1(A) = \max\left(\min\{0.032n^4, 16\lambda^{-1}\}, 3\right). \tag{3.3}$$

See Fig. 3.1 for a comparison of $\kappa(A)$ and $\kappa_1(A)$ when $n = 5000$. Clearly, A is very ill-conditioned when λ is very small (as it often is in practice). We will investigate whether a QR approach improves the RMS error of the spline.

3.2   Least-Squares Formulation and QR Factorization

With the Cholesky factorization

$$P = B^T B, \tag{3.4}$$

we can write

$$A = F^T F, \tag{3.5}$$

where

$$F = \begin{bmatrix} M^T \\ \sqrt{\lambda}B \end{bmatrix}. \tag{3.6}$$

Therefore, if

$$c = \arg\min_{\zeta} \left\| F\zeta - \begin{bmatrix} y \\ 0 \end{bmatrix} \right\|, \tag{3.7}$$

then c satisfies (1.7) and the spline can be obtained from (1.8). Suppose we can find an orthogonal matrix Q ($Q^T Q = I$) such that

$$F = Q \begin{bmatrix} R \\ 0 \end{bmatrix}, \tag{3.8}$$

where R is upper triangular. If

$$Q^T \begin{bmatrix} y \\ 0 \end{bmatrix} = \begin{bmatrix} r \\ b \end{bmatrix}, \tag{3.9}$$

then

$$\left\| F\zeta - \begin{bmatrix} y \\ 0 \end{bmatrix} \right\|^2 = \left\| Q^T F\zeta - Q^T \begin{bmatrix} y \\ 0 \end{bmatrix} \right\|^2 = \left\| \begin{bmatrix} R \\ 0 \end{bmatrix}\zeta - \begin{bmatrix} r \\ b \end{bmatrix} \right\|^2$$
$$= \|R\zeta - r\|^2 + \|b\|^2, \tag{3.10}$$

and c can be found by solving the triangular system

$$Rc = r. \tag{3.11}$$

Note that

$$A = F^T F = R^T R, \tag{3.12}$$

so R is a Cholesky factor of A and, like the normalized Cholesky factor L in (2.1), it is banded with bandwidth two. Since

$$R = D^{\frac{1}{2}}L^T, \tag{3.13}$$

r is related to θ in (2.9) via

$$r = D^{\frac{1}{2}}\theta. \tag{3.14}$$

We can construct Q as a product of $(4n - 9)$ Givens (plane) rotations [1]:

$$Q^T = V_{n-2}^T U_{n-2}^T \cdots V_1^T U_1^T, \tag{3.15}$$

where U_i^T, $1 \le i \le n - 3$, is the product of two Givens rotations which zero entries $(i + n, i)$ and $(i + n, i + 1)$ of F, U_{n-2}^T is a single Givens rotation which zeros entry $(2n - 2, n - 2)$, and V_i^T, $1 \le i \le n - 2$, is the product of two Givens rotations which zero the $(i + 1, i)$ and $(i + 2, i)$ entries while placing two nonzero numbers in positions $(i, i + 1)$ and $(i, i + 2)$. De Hoog and Hutchinson [1] do not store or use R, but instead store Q and compute c and s using alternatives to (3.11) and (1.8). Our algorithm will require less storage and execute more quickly. Note also that the Cholesky factorization of P can be carried out to within machine precision in just 14 iterations regardless of n [8].

To see how (3.15) produces R from F, let $n = 7$. Then

$$F = \begin{bmatrix} 1 & 0 & 0 & 0 & 0 \\ -2 & 1 & 0 & 0 & 0 \\ 1 & -2 & 1 & 0 & 0 \\ 0 & 1 & -2 & 1 & 0 \\ 0 & 0 & 1 & -2 & 1 \\ 0 & 0 & 0 & 1 & -2 \\ 0 & 0 & 0 & 0 & 1 \\ x & x & 0 & 0 & 0 \\ 0 & x & x & 0 & 0 \\ 0 & 0 & x & x & 0 \\ 0 & 0 & 0 & x & x \\ 0 & 0 & 0 & 0 & x \end{bmatrix} \rightarrow \begin{bmatrix} x & 0 & 0 & 0 & 0 \\ x & x & 0 & 0 & 0 \\ 1 & -2 & 1 & 0 & 0 \\ 0 & 1 & -2 & 1 & 0 \\ 0 & 0 & 1 & -2 & 1 \\ 0 & 0 & 0 & 1 & -2 \\ 0 & 0 & 0 & 0 & 1 \\ 0 & 0 & 0 & 0 & 0 \\ 0 & x & x & 0 & 0 \\ 0 & 0 & x & x & 0 \\ 0 & 0 & 0 & x & x \\ 0 & 0 & 0 & 0 & x \end{bmatrix} \rightarrow \begin{bmatrix} x & x & x & 0 & 0 \\ 0 & x & 0 & 0 & 0 \\ 0 & x & x & 0 & 0 \\ 0 & 1 & -2 & 1 & 0 \\ 0 & 0 & 1 & -2 & 1 \\ 0 & 0 & 0 & 1 & -2 \\ 0 & 0 & 0 & 0 & 1 \\ 0 & 0 & 0 & 0 & 0 \\ 0 & x & x & 0 & 0 \\ 0 & 0 & x & x & 0 \\ 0 & 0 & 0 & x & x \\ 0 & 0 & 0 & 0 & x \end{bmatrix}$$

$$\rightarrow \begin{bmatrix} x & x & x & 0 & 0 \\ 0 & x & 0 & 0 & 0 \\ 0 & x & x & 0 & 0 \\ 0 & 1 & -2 & 1 & 0 \\ 0 & 0 & 1 & -2 & 1 \\ 0 & 0 & 0 & 1 & -2 \\ 0 & 0 & 0 & 0 & 1 \\ 0 & 0 & 0 & 0 & 0 \\ 0 & 0 & 0 & 0 & 0 \\ 0 & 0 & x & x & 0 \\ 0 & 0 & 0 & x & x \\ 0 & 0 & 0 & 0 & x \end{bmatrix} \rightarrow \begin{bmatrix} x & x & x & 0 & 0 \\ 0 & x & x & x & 0 \\ 0 & 0 & x & 0 & 0 \\ 0 & 0 & x & x & 0 \\ 0 & 0 & 1 & -2 & 1 \\ 0 & 0 & 0 & 1 & -2 \\ 0 & 0 & 0 & 0 & 1 \\ 0 & 0 & 0 & 0 & 0 \\ 0 & 0 & 0 & 0 & 0 \\ 0 & 0 & x & x & 0 \\ 0 & 0 & 0 & x & x \\ 0 & 0 & 0 & 0 & x \end{bmatrix} \rightarrow \begin{bmatrix} x & x & x & 0 & 0 \\ 0 & x & x & x & 0 \\ 0 & 0 & x & 0 & 0 \\ 0 & 0 & x & x & 0 \\ 0 & 0 & 1 & -2 & 1 \\ 0 & 0 & 0 & 1 & -2 \\ 0 & 0 & 0 & 0 & 1 \\ 0 & 0 & 0 & 0 & 0 \\ 0 & 0 & 0 & 0 & 0 \\ 0 & 0 & 0 & 0 & 0 \\ 0 & 0 & 0 & x & x \\ 0 & 0 & 0 & 0 & x \end{bmatrix} \rightarrow \begin{bmatrix} x & x & x & 0 & 0 \\ 0 & x & x & x & 0 \\ 0 & 0 & x & x & x \\ 0 & 0 & 0 & x & 0 \\ 0 & 0 & 0 & x & x \\ 0 & 0 & 0 & 1 & -2 \\ 0 & 0 & 0 & 0 & 1 \\ 0 & 0 & 0 & 0 & 0 \\ 0 & 0 & 0 & 0 & 0 \\ 0 & 0 & 0 & 0 & 0 \\ 0 & 0 & 0 & x & x \\ 0 & 0 & 0 & 0 & x \end{bmatrix}$$

$$
\rightarrow
\begin{bmatrix}
x & x & x & 0 & 0 \\
0 & x & x & x & 0 \\
0 & 0 & x & x & x \\
0 & 0 & 0 & x & 0 \\
0 & 0 & 0 & x & x \\
0 & 0 & 0 & 1 & -2 \\
0 & 0 & 0 & 0 & 1 \\
0 & 0 & 0 & 0 & 0 \\
0 & 0 & 0 & 0 & 0 \\
0 & 0 & 0 & 0 & 0 \\
0 & 0 & 0 & 0 & 0 \\
0 & 0 & 0 & 0 & x
\end{bmatrix}
\rightarrow
\begin{bmatrix}
x & x & x & 0 & 0 \\
0 & x & x & x & 0 \\
0 & 0 & x & x & x \\
0 & 0 & 0 & x & x \\
0 & 0 & 0 & 0 & x \\
0 & 0 & 0 & 0 & x \\
0 & 0 & 0 & 0 & 1 \\
0 & 0 & 0 & 0 & 0 \\
0 & 0 & 0 & 0 & 0 \\
0 & 0 & 0 & 0 & 0 \\
0 & 0 & 0 & 0 & 0 \\
0 & 0 & 0 & 0 & x
\end{bmatrix}
\rightarrow
\begin{bmatrix}
x & x & x & 0 & 0 \\
0 & x & x & x & 0 \\
0 & 0 & x & x & x \\
0 & 0 & 0 & x & x \\
0 & 0 & 0 & 0 & x \\
0 & 0 & 0 & 0 & x \\
0 & 0 & 0 & 0 & 1 \\
0 & 0 & 0 & 0 & 0 \\
0 & 0 & 0 & 0 & 0 \\
0 & 0 & 0 & 0 & 0 \\
0 & 0 & 0 & 0 & 0 \\
0 & 0 & 0 & 0 & 0
\end{bmatrix}
\rightarrow
\begin{bmatrix}
x & x & x & 0 & 0 \\
0 & x & x & x & 0 \\
0 & 0 & x & x & x \\
0 & 0 & 0 & x & x \\
0 & 0 & 0 & 0 & x \\
0 & 0 & 0 & 0 & 0 \\
0 & 0 & 0 & 0 & 0 \\
0 & 0 & 0 & 0 & 0 \\
0 & 0 & 0 & 0 & 0 \\
0 & 0 & 0 & 0 & 0 \\
0 & 0 & 0 & 0 & 0 \\
0 & 0 & 0 & 0 & 0
\end{bmatrix}
=
\begin{bmatrix}
R \\
0
\end{bmatrix}.
$$

3.3 MATLAB Implementation

The QR algorithm is implemented in the MATLAB function *splineqr*. To keep the comparison with the Cholesky algorithm as simple as possible, no iteration truncation was used even though R exhibits the same convergence along diagonals as L. Therefore, *splineqr* will be compared to the function *csplineopt* given in Section 2.3. Both functions have the same general structure and use the same quantities to compute the GCV score.

```
function s = splineqr(y)
%
n = length(y);
nc = ceil(n/2);
rmndr = rem(n,2);
nn = min(n-2,15);
sig = fminbnd(@gcvqr, 0, 1);
gcvqr(sig);
s = y-s;
%
  function score = gcvqr(sig)
%
  s = y;
  f = zeros(1,nn);
  e = zeros(1,nn-1);
  lam = 4*sig^4/(1-sig^2);
  lamrt = sqrt(lam);
  f(1) = sqrt(2/3);
  for k = 2:nn
    e(k-1) = 1/(6*f(k-1));
    f(k) = sqrt(2/3-e(k-1)^2);
  end
  f = f*lamrt;
```

```
e = e*lamrt;
f = [f,f(nn)*ones(1,n-17)];
e = [e,e(nn-1)*ones(1,n-17)];
tm1 = -2;
tm2 = 1;
tm3 = 1;
for k = 1:n-3
  a = tm3;
  b = e(k);
  if abs(b) > abs(a)
    tg = a/b;
    tt = sqrt(1+tg^2);
    sn = 1/tt;
    cn = tg*sn;
    tm3 = tt*b;
  else
    tg = b/a;
    tt = sqrt(1+tg^2);
    cn = 1/tt;
    sn = tg*cn;
    tm3 = tt*a;
  end
  tm4 = cn*f(k)-sn*tm1;
  tm1 = cn*tm1+sn*f(k);
  t2 = -sn*s(k+1);
  s(k+1) = cn*s(k+1);
  a = tm2;
  b = tm4;
  if abs(b) > abs(a)
    tg = a/b;
    tt = sqrt(1+tg^2);
    sn = 1/tt;
    cn = tg*sn;
    tm2 = tt*b;
  else
    tg = b/a;
    tt = sqrt(1+tg^2);
    cn = 1/tt;
    sn = tg*cn;
    tm2 = tt*a;
  end
  s(k) = cn*s(k)+sn*t2;
  a = tm2;
  b = tm1;
  if abs(b) > abs(a)
```

```
      tg = a/b;
      tt = sqrt(1+tg^2);
      sn = 1/tt;
      cn = tg*sn;
      tm2 = tt*b;
   else
      tg = b/a;
      tt = sqrt(1+tg^2);
      cn = 1/tt;
      sn = tg*cn;
      tm2 = tt*a;
   end
   tm1 = sn*tm3;
   tm3 = cn*tm3;
   t1 = cn*s(k)+sn*s(k+1);
   s(k+1) = -sn*s(k)+cn*s(k+1);
   s(k) = t1;
   a = tm2;
   if abs(a) < 1
      tt = sqrt(1+a^2);
      sn = 1/tt;
      cn = a*sn;
      f(k) = tt;
   else
      tg = 1/a;
      tt = sqrt(1+tg^2);
      cn = 1/tt;
      sn = tg*cn;
      f(k) = tt*a;
   end
   e(k) = (-cn*tm1+2*sn)/f(k);
   tm1 = -2*cn-sn*tm1;
   if k < n-3
      tm2 = tm3;
      tm3 = cn;
   else
   end
   t1 = cn*s(k)+sn*s(k+2);
   s(k+2) = -sn*s(k)+cn*s(k+2);
   s(k) = t1/f(k);
   f(k) = 1/(f(k)^2);
end
a = tm3;
b = f(n-2);
if abs(b) > abs(a)
```

```
  tg = a/b;
  tt = sqrt(1+tg^2);
  cn = tg/tt;
  tm3 = tt*b;
else
  tg = b/a;
  tt = sqrt(1+tg^2);
  cn = 1/tt;
  tm3 = tt*a;
end
s(n-2) = cn*s(n-2);
a = tm3;
b = tm1;
if abs(b) > abs(a)
  tg = a/b;
  tt = sqrt(1+tg^2);
  sn = 1/tt;
  cn = tg*sn;
  tm3 = tt*b;
else
  tg = b/a;
  tt = sqrt(1+tg^2);
  cn = 1/tt;
  sn = tg*cn;
  tm3 = tt*a;
end
s(n-2) = cn*s(n-2)+sn*s(n-1);
a = tm3;
if abs(a) < 1
  tt = sqrt(1+a^2);
  sn = 1/tt;
  cn = a*sn;
  f(n-2) = tt;
else
  tg = 1/a;
  tt = sqrt(1+tg^2);
  cn = 1/tt;
  sn = tg*cn;
  f(n-2) = tt*a;
end
s(n-2) = (cn*s(n-2)+sn*s(n))/f(n-2);
f(n-2) = 1/(f(n-2)^2);
s = s(1:n-2);
s(n-3) = s(n-3)+e(n-3)*s(n-2);
for k = n-4:-1:1
```

Table 3.1 RMS error for $n = 10^6$

signal	SNR=20 dB *csplineopt*	SNR=20 dB *splineqr*	SNR=40 dB *csplineopt*	SNR=40 dB *splineqr*
x_1	1.7×10^{-2}	1.7×10^{-2}	2.2×10^{-3}	2.2×10^{-3}
x_2	6.1×10^{-3}	7.0×10^{-3}	6.3×10^{-4}	7.3×10^{-4}
x_3	8.9×10^{-3}	1.0×10^{-2}	9.0×10^{-4}	1.1×10^{-3}

```
      s(k) = s(k)+e(k)*s(k+1)-f(k)*s(k+2);
    end
    g2 = f(n-2);
    d = e(n-3)*g2;
    tr2 = d;
    g1 = f(n-3)+e(n-3)*d;
    tr1 = g2+g1;
    tr3 = 0;
    for j = n-4:-1:n-nc
      p = e(j)*d-f(j)*g2;
      tr3 = tr3+p;
      d = e(j)*g1-f(j)*d;
      tr2 = tr2+d;
      g2 = g1;
      g1 = f(j)*(1-p)+e(j)*d;
      tr1 = tr1+g1;
    end
    p = e(n-nc-1)*d-f(n-nc-1)*g2;
    d = e(n-nc-1)*g1-f(n-nc-1)*d;
    tr = 6*(2*tr1-rmndr*g1)-8*(2*tr2+(1-rmndr)*d)...
            +2*(2*tr3+(2-rmndr)*p);
    s = diff([0; 0; s; 0; 0],2);
    score = n*(s'*s)/tr^2;
    end
end
```

For large n, if m values of σ are used to minimize the GCV score, *splineqr* requires about $n + 63.5mn$ additions and multiplications, $11mn$ divisions, and $4mn$ square roots, whereas *csplineopt* requires about $3n + 23.5mn$ additions and multiplications and mn divisions. Consequently, we expect *splineqr* to be significantly slower than *csplineopt*. On the other hand, *splineqr* uses $32n$ bytes of memory compared to $40n$ bytes for *csplineopt*.

3.4 Monte Carlo Simulations

The simulation regime described in Section 2.4 was used to compare *splineqr* to *csplineopt* in terms of speed and accuracy. For all cases, *csplineopt* was about four times faster than *splineqr*. For the second and third signals with $n = 10^6$, *csplineopt* produced a smaller RMS error than *splineqr*. For all other cases, the RMS errors were the same. See Table 3.1. So other than a 20% reduction in memory use, the QR algorithm offers no advantage over the Cholesky algorithm.

References

[1] De Hoog FR, Hutchinson MF (1987) An efficient method for calculating smoothing splines using orthogonal transformations. Numer Math 50:311-319
[2] George A, Heath MT (1980) Solution of sparse linear least squares problems using Givens rotations. Linear Algebra Appl 34:69-83
[3] Cox MG (1981) The least squares solution of overdetermined linear equations having band or augmented band structure. IMA J Numer Anal 1:3-22
[4] Elden L (1984) An algorithm for the regularization of ill-conditioned banded least squares problems. SIAM J Sci Stat Comput 5:237-254
[5] Golub GH, Van Loan CF (1996) Matrix computations. The Johns Hopkins University Press, Baltimore
[6] Stewart GW (1998) Matrix algorithms - basic decompositions. SIAM, Philadelphia
[7] Higham NJ (2002) Accuracy and stability of numerical algorithms. SIAM, Philadelphia
[8] Malcolm MA, Palmer J (1974) A fast method for solving a class of tridiagonal linear systems. Commun. ACM 17:14-17

Chapter 4
FFT Algorithm

For a given λ, the cubic spline smoother is a time-varying linear filter, but it can be approximated by a time-invariant linear filter. It will then be amenable to frequency domain analysis and implementation using the FFT. The FFT algorithm will be compared to the Cholesky algorithm in terms of execution time, accuracy, and memory use. For digital signal processing background, see [1]. Other results on splines in the frequency domain can be found in [2].

4.1   Frequency Response of the Spline Smoother

Eq. (1.7) can be viewed as a fourth-order difference equation

$$c_i + (\lambda/6 - 4)c_{i+1} + (2\lambda/3 + 6)c_{i+2} + (\lambda/6 - 4)c_{i+3} + c_{i+4}$$
$$= y_{i+2} - 2y_{i+3} + y_{i+4}, \tag{4.1}$$

for $1 \leq i \leq n - 6$, with left boundary conditions

$$(2\lambda/3 + 6)c_1 + (\lambda/6 - 4)c_2 + c_3 = y_1 - 2y_2 + y_3,$$
$$(\lambda/6 - 4)c_1 + (2\lambda/3 + 6)c_2 + (\lambda/6 - 4)c_3 + c_4 = y_2 - 2y_3 + y_4, \tag{4.2}$$

and right boundary conditions

$$c_{n-5} + (\lambda/6 - 4)c_{n-4} + (2\lambda/3 + 6)c_{n-3} + (\lambda/6 - 4)c_{n-2} = y_{n-3} - 2y_{n-2} + y_{n-1},$$
$$c_{n-4} + (\lambda/6 - 4)c_{n-3} + (2\lambda/3 + 6)c_{n-2} = y_{n-2} - 2y_{n-1} + y_n. \tag{4.3}$$

Suppose we ignore the boundary conditions and consider (4.1) to be valid for all i. We will also ignore the first two and last two equations in (1.8) and use

H.L. Weinert, *Fast Compact Algorithms and Software for Spline Smoothing*, 29
SpringerBriefs in Computer Science, DOI 10.1007/978-1-4614-5496-0_4,
© The Author(s) 2013

$$s_i = y_i - c_{i-2} + 2c_{i-1} - c_i, \qquad (4.4)$$

for all i. Taking bilateral z-transforms of (4.1) and (4.4),

$$\left(z^4 + (\lambda/6 - 4)z^3 + (2\lambda/3 + 6)z^2 + (\lambda/6 - 4)z + 1\right)C(z) = z^2(z-1)^2 Y(z),$$
$$S(z) = Y(z) - z^{-2}(z-1)^2 C(z). \qquad (4.5)$$

The transfer function $H(z)$ of the spline smoother is thus

$$H(z) = \frac{S(z)}{Y(z)} = \frac{\lambda z^2 (z + 4 + z^{-1})}{6(z-1)^4 + \lambda z^2(z + 4 + z^{-1})}. \qquad (4.6)$$

The denominator polynomial is a scaled version of (2.15). Since there are no poles (denominator roots) on the unit circle, a frequency response $H(\omega)$ exists and can be obtained by replacing z in (4.6) with $e^{j\omega}$. After some simplification,

$$H(\omega) = \frac{\lambda(2 + \cos \omega)}{12(1 - \cos \omega)^2 + \lambda(2 + \cos \omega)}. \qquad (4.7)$$

This frequency response is real and periodic with period 2π. It is also even so we can restrict attention to the interval $\omega \in [0, \pi]$. Fig. 4.1 shows plots of $H(\omega)$ for various λ.

We see that $H(\omega)$ is nonnegative, achieves a maximum value of one at the origin, and decreases monotonically with increasing frequency (no ripples). Furthermore,

$$\lim_{\lambda \to \infty} H(\omega) = 1, \quad \forall \omega,$$

$$\lim_{\lambda \to 0} H(\omega) = \begin{cases} 1, & \omega = 0 \\ 0, & \omega \neq 0 \end{cases}, \qquad (4.8)$$

$$H(\pi) = \frac{\lambda}{\lambda + 48}. \qquad (4.9)$$

For small ω,

$$H(\omega) \cong 1 - \lambda^{-1}\omega^4. \qquad (4.10)$$

which implies that the first three derivatives of $H(\omega)$ are zero at the origin. Consequently, the smoother passes cubic polynomials unchanged.

The 3 dB cutoff frequency (or bandwidth) in pi units, plotted in Fig. 4.2, is

$$\frac{\omega_0}{\pi} = \pi^{-1}\cos^{-1}\left(1 + \tfrac{1}{24}\left(\eta\lambda - \sqrt{\eta\lambda(\eta\lambda + 144)}\right)\right), \quad \eta = \sqrt{2} - 1, \qquad (4.11)$$

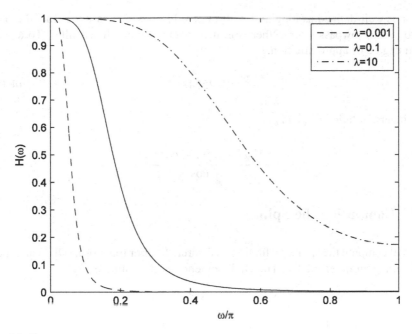

Fig. 4.1 Frequency response

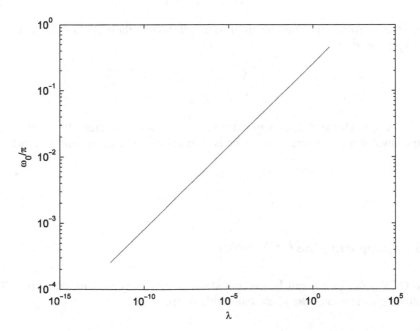

Fig. 4.2 Bandwidth

which is well-defined as long as $\lambda \leq 48(\sqrt{2}+1)$, in which case the smoother is a (zero-phase) lowpass filter. Otherwise, it is essentially an allpass filter. To a very good degree of approximation,

$$\frac{\omega_0}{\pi} \cong 0.2554\lambda^{1/4}. \tag{4.12}$$

The inverse relation to (4.11) is

$$\lambda = \frac{12(\sqrt{2}+1)(1-\cos\omega_0)^2}{2+\cos\omega_0}. \tag{4.13}$$

4.2 Computing the Spline

We will compute the spline by finding its discrete Fourier transform (DFT) and then performing an inverse DFT. The DFT sequence of the spline is

$$S_k = H_k Y_k, \quad 1 \leq k \leq n, \tag{4.14}$$

where Y_k is the DFT sequence of the measurements and

$$H_k = H((k-1)2\pi/n). \tag{4.15}$$

Since the measurements and the spline are real-valued, their DFTs exhibit conjugate symmetry:

$$\begin{aligned}
S_k &= S^*_{n-k+2}, \\
Y_k &= Y^*_{n-k+2},
\end{aligned} \tag{4.16}$$

for $2 \leq k \leq n$. This symmetry will reduce execution time and memory use since all frequency domain computations can be performed with vectors of length n_0, where

$$n_0 = \text{floor}(n/2) + 1. \tag{4.17}$$

4.3 Computing the GCV Score

The GCV score (1.10) can be computed using frequency domain quantities. To evaluate the numerator use (1.8), Parseval's relation, and (4.14):

$$n^{-1}c^T M M^T c = n^{-1}\sum_{i=1}^{n}(y_i - s_i)^2 = n^{-2}\sum_{i=1}^{n}|Y_i - S_i|^2 = n^{-2}\sum_{i=1}^{n}|(1-H_i)Y_i|^2. \tag{4.18}$$

In view of (4.16) and the fact that $H_1 = 1$,

$$n^{-1}c^T M M^T c = n^{-2}\left(2\sum_{i=2}^{n_0}|(1-H_i)Y_i|^2 - \gamma|(1-H_{n_0})Y_{n_0}|^2\right). \qquad (4.19)$$

where

$$\gamma = \begin{cases} 1, & n \text{ even} \\ 0, & n \text{ odd} \end{cases}. \qquad (4.20)$$

The matrix $M^T A^{-1} M$ in the denominator of (1.10) maps y to $y - s$. With the approximations discussed in Section 4.1, the spline can be written as the convolution

$$s_i = \sum_{k=1}^{n} h_{i-k}y_k, \qquad (4.21)$$

where the impulse response sequence h_i is the inverse Fourier transform of the frequency response $H(\omega)$. The impulse response sequence has infinite extent, and it is real-valued and even. In matrix form, (4.21) becomes

$$s = \begin{bmatrix} h_0 & h_1 & h_2 & \cdots & h_{n-1} \\ h_1 & h_0 & h_1 & \ddots & \vdots \\ h_2 & h_1 & h_0 & \ddots & h_2 \\ \vdots & \ddots & \ddots & \ddots & h_1 \\ h_{n-1} & \cdots & h_2 & h_1 & h_0 \end{bmatrix} y. \qquad (4.22)$$

Hence,

$$n^{-1}\text{trace}\left(M^T A^{-1} M\right) = 1 - h_0 = 1 - (2\pi)^{-1}\int_0^{2\pi} H(\omega)\,d\omega. \qquad (4.23)$$

Since

$$H_k = H_{n-k+2}, \qquad (4.24)$$

for $2 \le k \le n$, the trapezoidal integration method gives

$$n^{-1}\text{trace}\left(M^T A^{-1} M\right) \cong n^{-1}\left(2\sum_{i=2}^{n_0}(1 - H_i) - \gamma(1 - H_{n_0})\right). \tag{4.25}$$

To summarize,

$$GCV(\lambda) \cong \frac{2\sum_{i=2}^{n_0}\left|(1 - H_i)Y_i\right|^2 - \gamma\left|(1 - H_{n_0})Y_{n_0}\right|^2}{\left(2\sum_{i=2}^{n_0}(1 - H_i) - \gamma(1 - H_{n_0})\right)^2}. \tag{4.26}$$

An exact formula for the GCV score appears (modulo a typographical error) in [3], but the formula is numerically unreliable when λ is near 24.

4.4 MATLAB Implementation

The FFT algorithm is implemented in the MATLAB function *splinefft*. The input is the column vector of measurements *y*. The output is the column vector of spline values *s*. The minimization of the GCV score is carried out by the MATLAB function *fminbnd*. All computations that must be repeated when the smoothing parameter is changed are in the nested function *gcvfft*. After *fminbnd* terminates, an extra call to *gcvfft* is necessary to compute, using the optimal σ, those quantities that determine the spline. The total memory requirement is $28n$ bytes, compared to $24n$ or $40n$ for *splinechol*.

```
function s = splinefft(y)
%
n = length(y);
cw = cos(0:2*pi/n:pi)';
cw = (2+cw)./(12*((1-cw).^2));
s = fft(y);
s = s(1:length(cw));
gamma = ~mod(n,2);
sig = fminbnd(@gcvfft, 0, 1);
gcvfft(sig);
m = floor((n+1)/2);
s = s-HE;
s = ifft([s;conj(s(m:-1:2))]);
%
    function score = gcvfft(sig)
%
    lam = 4*sig^4/(1-sig^2);
    HE = 1./(1+lam*cw); % 1-H
    tr = 2*sum(HE)-gamma*HE(end);
```

Table 4.1 Execution time (sec.) for $n = 2^{20} = 1{,}048{,}576$

signal	*splinechol*	*splinefft*
x_1	1.1	0.23
x_2	1.4	0.27
x_3	1.5	0.28

Table 4.2 Execution time (sec.) for $n = 2^{23} = 8{,}388{,}608$

signal	*splinechol*	*splinefft*
x_1	6.1	1.3
x_2	6.4	1.3
x_3	6.4	1.3

```
HE = s.*HE; % Y-S
score = (2*(HE'*HE)-gamma*abs(HE(end))^2)/tr^2;
end
end
```

4.5 Monte Carlo Simulations

The estimation accuracy of the time-domain function *splinechol* and the frequency-domain function *splinefft* were compared for the same three signals described in Section 2.4. Interestingly, the approximations used in *splinefft* do not significantly affect the RMSE. Sometimes *splinechol* is a little more accurate, and sometimes thereverse is true. As for execution time, *splinefft* is never slower than *splinechol*. When n is a large power of two, *splinefft* is about five times faster than *splinechol*, independently of signal type or SNR. Tables 4.1 and 4.2 show examples. When n is not a power of two, the speed advantage of *splinefft* is not as great.

In summary, if one compares the Cholesky, QR, and FFT algorithms for continuous spline smoothing on the basis of speed, memory use, and estimation accuracy, it is clear that the FFT algorithm is the best option.

References

[1] Manolakis DG, Ingle VK (2011) Applied digital signal processing. Cambridge University Press, New York
[2] Unser M, Blu T (2007) Self-similarity: part 1-splines and operators. IEEE Trans Signal Process 55:1352–1363
[3] De Nicolao G, Ferrari-Trecate G, Sparacino G (2000) Fast spline smoothing via spectral factorization concepts. Automatica 36:1733–1739

Chapter 5
Discrete Spline Smoothing

In this chapter we investigate Cholesky and FFT algorithms for discrete cubic spline smoothing, and we compare the performance of the continuous and discrete smoothers. Henderson [1] was the first to use Cholesky factorization for the discrete problem.

5.1 The Discrete Problem

In the discrete problem, the integral in the cost functional (1.2) is replaced by the sum of squared second differences:

$$\xi \sum_{i=1}^{n} (y_i - u(i/n))^2 + \sum_{i=1}^{n-2} (u((i+2)/n) - 2u((i+1)/n) + u(i/n))^2. \qquad (5.1)$$

In matrix form this becomes

$$\xi(y-u)^T(y-u) + u^T M^T M u. \qquad (5.2)$$

The minimizer s of (5.2) satisfies

$$Ac = \left(\xi I + MM^T\right)c = My, \qquad (5.3)$$

$$s = y - M^T c. \qquad (5.4)$$

H.L. Weinert, *Fast Compact Algorithms and Software for Spline Smoothing*,
SpringerBriefs in Computer Science, DOI 10.1007/978-1-4614-5496-0_5,
© The Author(s) 2013

When $n = 8$ for example,

$$A = \begin{bmatrix} \xi+6 & -4 & 1 & 0 & 0 & 0 \\ -4 & \xi+6 & -4 & 1 & 0 & 0 \\ 1 & -4 & \xi+6 & -4 & 1 & 0 \\ 0 & 1 & -4 & \xi+6 & -4 & 1 \\ 0 & 0 & 1 & -4 & \xi+6 & -4 \\ 0 & 0 & 0 & 1 & -4 & \xi+6 \end{bmatrix}. \tag{5.5}$$

As in the continuous case, A is positive definite, Toeplitz, and pentadiagonal.

5.2   Cholesky Algorithm

The Cholesky algorithm (2.5)-(2.14) is valid for the discrete case as long as the first equation in (2.7) is changed to

$$a_0 = \xi + 6, \quad a_1 = 4. \tag{5.6}$$

The polynomial whose roots determine the number of iterations required for convergence is now

$$z^4 - 4z^3 + (\xi + 6)z^2 - 4z + 1. \tag{5.7}$$

Eq. (2.17) is still valid, but (2.18) must be modified slightly:

$$N_1 = \begin{cases} \text{ceil}\left(26\xi^{-1/4}\right), & \xi \le 100 \\ 8, & \xi > 100 \end{cases}. \tag{5.8}$$

See Fig. 5.1 and Table 5.1. The GCV score can be computed exactly as detailed in Section 2.2.

The MATLAB function *dsplinechol* implements the discrete Cholesky algorithm.

```
function s = dsplinechol(y)
%
n = length(y);
nc = ceil(n/2);
rmndr = rem(n,2);
w = diff(y,2);
sig = fminbnd(@dgcv, 0, 1);
dgcv(sig);
```

Table 5.1 Number of
iterations for convergence

ξ	10^{-12}	10^{-10}	10^{-8}	10^{-6}	10^{-4}	10^{-2}	1	100
N	25977	8215	2598	822	260	82	26	8
N_1	26000	8222	2600	823	260	83	26	9

```
s = y-s;
%
  function score = dgcv(sig)
%
  s = zeros(n-2,1);
  xi = 4*sig^4/(1-sig^2);
  a0 = 6+xi;
  if xi > 100
    N = 8;
  else
    N = ceil(26*xi^-.25);
  end
  if N > n-5 % No truncation
      e = zeros(1,n-2);
      f = zeros(1,n-2);
      f(1) = 1/a0;
      s(1) = f(1)*w(1);
      e(1) = 4*f(1);
      f(2) = 1/(a0-4*e(1));
      s(2) = f(2)*(w(2)+4*s(1));
      mu = 4-e(1);
      e(2) = mu*f(2);
      for k = 3:n-2
        f(k) = 1/(a0-mu*e(k-1)-f(k-2));
        s(k) = f(k)*(w(k)+mu*s(k-1)-s(k-2));
        mu = 4-e(k-1);
        e(k) = mu*f(k);
      end
      s(n-3) = s(n-3)+e(n-3)*s(n-2);
      for k = n-4:-1:1
          s(k) = s(k)+e(k)*s(k+1)-f(k)*s(k+2);
      end
      g2 = f(n-2);
      tr1 = g2;
      d = e(n-3)*g2;
      tr2 = d;
      g1 = f(n-3)+e(n-3)*d;
      tr1 = tr1+g1;
      tr3 = 0;
      for k = n-4:-1:n-nc
```

```
           p = e(k)*d-f(k)*g2;
           tr3 = tr3+p;
           d = e(k)*g1-f(k)*d;
           tr2 = tr2+d;
           g2 = g1;
           g1 = f(k)*(1-p)+e(k)*d;
           tr1 = tr1+g1;
       end
       p = e(n-nc-1)*d-f(n-nc-1)*g2;
       tr3 = tr3+p;
       d = e(n-nc-1)*g1-f(n-nc-1)*d;
       tr2 = tr2+d;
   else % Truncate e, f iterations
       e = zeros(1,N);
       f = zeros(1,N);
       f(1) = 1/a0;
       s(1) = f(1)*w(1);
       e(1) = 4*f(1);
       f(2) = 1/(a0-4*e(1));
       s(2) = f(2)*(w(2)+4*s(1));
       mu = 4-e(1);
       e(2) = mu*f(2);
       for k = 3:N
         f(k) = 1/(a0-mu*e(k-1)-f(k-2));
         s(k) = f(k)*(w(k)+mu*s(k-1)-s(k-2));
         mu = 4-e(k-1);
         e(k) = mu*f(k);
       end
       flim = f(N);
       elim = e(N);
       mu = 4-elim;
       for k = N+1:n-2
         s(k) = flim*(w(k)+mu*s(k-1)-s(k-2));
       end
       s(n-3) = s(n-3)+elim*s(n-2);
       for k = n-4:-1:N
         s(k) = s(k)+elim*s(k+1)-flim*s(k+2);
       end
       for k = N-1:-1:1
         s(k) = s(k)+e(k)*s(k+1)-f(k)*s(k+2);
       end
       g2 = flim;
       tr1 = g2;
       d = elim*g2;
       tr2 = d;
       g1 = flim+elim*d;
```

```
        tr1 = tr1+g1;
        tr3 = 0;
    if N < nc-1 % Truncate g,d,p iterations
        for k = 3:N
            p = elim*d-flim*g2;
            tr3 = tr3+p;
            d = elim*g1-flim*d;
            tr2 = tr2+d;
            g2 = g1;
            g1 = flim*(1-p)+elim*d;
            tr1 = tr1+g1;
        end
        tr1 = tr1+(nc-N-1)*g1;
        tr2 = tr2+(nc-N)*d;
        tr3 = tr3+(nc-N)*p;
    else % Don't truncate g,d,p iterations
        for k = n-4:-1:N
            p = elim*d-flim*g2;
            tr3 = tr3+p;
            d = elim*g1-flim*d;
            tr2 = tr2+d;
            g2 = g1;
            g1 = flim*(1-p)+elim*d;
            tr1 = tr1+g1;
        end
        for k = N-1:-1:n-nc
            p = e(k)*d-f(k)*g2;
            tr3 = tr3+p;
            d = e(k)*g1-f(k)*d;
            tr2 = tr2+d;
            g2 = g1;
            g1 = f(k)*(1-p)+e(k)*d;
            tr1 = tr1+g1;
        end
        p = e(n-nc-1)*d-f(n-nc-1)*g2;
        tr3 = tr3+p;
        d = e(n-nc-1)*g1-f(n-nc-1)*d;
        tr2 = tr2+d;
    end
end
tr = 6*(2*tr1-rmndr*g1)-8*(2*tr2-(1+rmndr)*d)...
    +2*(2*tr3-rmndr*p);
s = diff([0; 0; s; 0; 0],2);
score = n*(s'*s)/tr^2;
end
end
```

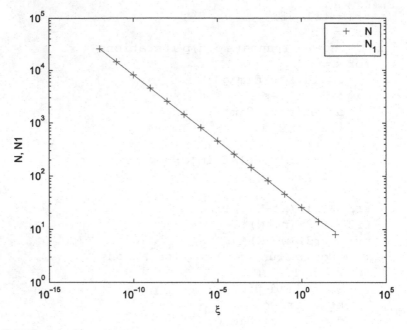

Fig. 5.1 Number of iterations for convergence

5.3   FFT Algorithm

Starting with (5.3)-(5.4) and proceeding as in Section 4.1, we find that the discrete spline smoother has frequency response (see Fig. 5.2)

$$H(\omega) = \frac{\xi}{4(1 - \cos\omega)^2 + \xi}. \tag{5.9}$$

This frequency response has all the properties of the continuous case except that

$$H(\pi) = \frac{\xi}{\xi + 16}, \tag{5.10}$$

and the cutoff frequency in pi units is now (see Fig. 5.3)

$$\frac{\omega_0}{\pi} = \pi^{-1}\cos^{-1}\left(1 - \tfrac{1}{2}\sqrt{\eta\xi}\right), \quad \eta = \sqrt{2} - 1, \tag{5.11}$$

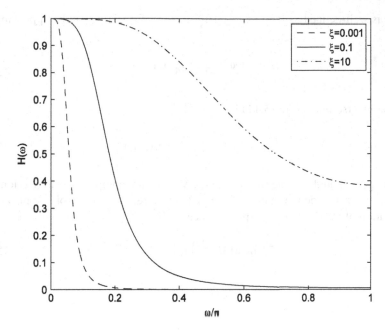

Fig. 5.2 Frequency response

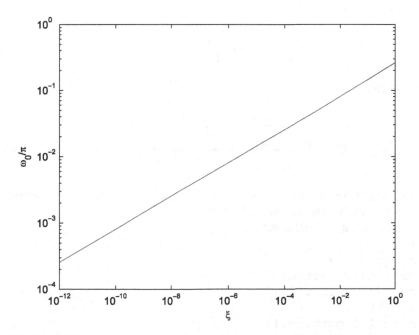

Fig. 5.3 Bandwidth

which is well-defined as long as $\xi \leq 16(\sqrt{2}+1)$. Eq. (5.11) can be approximated by

$$\frac{\omega_0}{\pi} \cong 0.2554\xi^{1/4}, \tag{5.12}$$

and the inverse relation to (5.11) is

$$\xi = 4\left(\sqrt{2}+1\right)(1-\cos\omega_0)^2. \tag{5.13}$$

The spline and the numerator of the GCV score are computed as in Sections 4.2 and 4.3. For the denominator of the GCV score, we can avoid approximate integration of the frequency response since

$$n^{-1}\mathrm{trace}\left(M^T A^{-1} M\right) = 1 - h_0, \tag{5.14}$$

and [2]

$$h_0 = \frac{\sigma}{2-\sigma^2}, \tag{5.15}$$

where σ and ξ are related by

$$\xi = \frac{4\sigma^4}{1-\sigma^2}. \tag{5.16}$$

Consequently,

$$GCV(\xi) = \frac{2\sum_{i=2}^{n_0}|(1-H_i)Y_i|^2 - \gamma|(1-H_{n_0})Y_{n_0}|^2}{(n(1-h_0))^2}. \tag{5.17}$$

The MATLAB function *dsplinefft* implements the discrete FFT algorithm. The total memory requirement is $28n$ bytes.

```
function s = dsplinefft(y)
%
n = length(y);
cw = cos(0:2*pi/n:pi)';
cw = 4*((1-cw).^2);
s = fft(y);
s = s(1:length(cw));
gamma = ~mod(n,2);
sig = fminbnd(@gcvfftd, 0, 1);
gcvfftd(sig);
```

```
m = floor((n+1)/2);
s = s-E; % Y-(Y-S)
s = ifft([s;conj(s(m:-1:2))]);
%
   function score = gcvfftd(sig)
%
   xi = 4*sig^4/(1-sig^2);
   E = s.*(cw./(xi+cw)); % Y-S
   tr = 1-sig/(2-sig^2);
   score = (2*(E'*E)-gamma*abs(E(end))^2)/(n*tr)^2;
   end
end
```

5.4 Discrete Versus Continuous

As in the continuous case, *dsplinefft* is about five times faster than *dsplinechol* when n is a large power of two, and they have roughly the same estimation accuracy and memory use. The remaining question is which FFT algorithm is better. Simulations show that *dsplinefft* and *splinefft* have the same accuracy and memory use, but *dsplinefft* is about 10% faster and is thus the best algorithm for cubic spline smoothing. By using a different approximation to the discrete smoothing spline, Garcia [3] developed a frequency domain algorithm based on the discrete cosine transform. While it is just as accurate as *dsplinefft*, it takes three times longer to execute.

References

[1] Henderson R (1925) Further remarks on graduation. Trans Actuar Soc Am 26:52–57
[2] Weinert HL (2007) Efficient computation for Whittaker-Henderson smoothing. Comput Stat Data Anal 52:959–974
[3] Garcia D (2010) Robust smoothing of gridded data in one and higher dimensions with missing values. Comput Stat Data Anal 54:1167–1178